Charlyan de Sousa Lima

Protein and amino acid supplementation in tambaqui feed

Charlyan de Sousa Lima

Protein and amino acid supplementation in tambaqui feed

Fish nutrition studies with a focus on fingerling

ScienciaScripts

Imprint

Cover image: www.ingimage.com

This book is a translation from the original published under ISBN 978-613-9-60332-9.

Publisher:
Sciencia Scripts
is a trademark of
Dodo Books Indian Ocean Ltd. and OmniScriptum S.R.L publishing group

120 High Road, East Finchley, London, N2 9ED, United Kingdom
Str. Armeneasca 28/1, office 1, Chisinau MD-2012, Republic of Moldova, Europe
Printed at: see last page
ISBN: 978-620-7-27316-4

"Wisdom is the highest perfection of reason and its main function is to perceive order in things." (St Thomas Aquinas)

DEDICATE,

To GOD, the inexhaustible source of wisdom, To my family, for helping me learn.

SUMMARY

With the aim of evaluating the effects of crude protein levels and their reduction with amino acid supplementation, based on the ideal protein concept, in rations for tambaqui (*Colossoma macropomum*) fry, two experiments were carried out in the fish nutrition research sector of the Centre for Agricultural and Environmental Sciences at the Federal University of Maranhão. In both experiments, the fry were kept in 30 polyethylene aquariums with a volumetric capacity of 500 litres in a recirculation system. The experimental diets were fed in five meals a day for 45 days and performance variables, carcass yield, body moisture, fat and protein content, body protein and fat deposition and nitrogen retention efficiency were assessed. The first experiment aimed to evaluate the effects of crude protein (CP) levels for tambaqui fry. 750 fry were used, with an initial weight of 0.35 ± 0.09 g, distributed in a completely randomised design, with six treatments, five replications and twenty-five fish per experimental unit. The treatments consisted of six isoenergetic, isocaloric and isophosphoric experimental rations, with different levels of crude protein (20.0; 24.0; 28.0; 32.0; 36.0 and 40.0 %). There were no effects of crude protein levels on feed intake and carcass yield. Crude protein intake increased linearly and the protein efficiency ratio decreased linearly as CP levels increased. Weight gain, feed conversion and specific growth rate improved up to the estimated levels of 31.57, 28.90 and 31.12 per cent CP, respectively. Body moisture was not influenced by increasing levels of CP in the feed. There was a quadratic effect of crude protein levels on body fat, body fat deposition, body protein, body protein deposition and nitrogen retention efficiency, with a minimum estimated level for the first two of 26.55; 23.77; and a maximum estimated level for the others of 29.26; 32.50; 27.21% CP, respectively. It was concluded that the crude protein recommendation for tambaqui fry is 31.57%, which corresponds to a digestible energy (DE):CP ratio of 9.50 kcal DE/g CP, as it provides better weight gain. In the second experiment, the aim was to evaluate the effects of reducing CP with amino acid supplementation, based on the ideal protein concept, in rations for tambaqui fry. A total of 750 fry with an initial weight of 0.44 ± 0.14 g were used in a completely randomised design with six treatments, five replications and twenty-five fish per experimental unit. The treatments consisted of six isoenergetic, isocaloric, isophosphoric and isolysin experimental rations containing different levels of CP (32.0, 30.5, 29.0, 27.5, 26.0 and 24.5 per cent). As the CP content decreased from 27.5 per cent CP, there was an improvement in weight gain, specific growth rate, feed conversion and protein efficiency ratio. Feed consumption was higher in fish fed diets containing 26% CP, followed by 29.0 and 27.5% CP, while the other levels did not vary. CP consumption decreased with the use of lower CP diets and carcass yield was not influenced by the treatments. Reducing the protein content of the feed did not influence body composition, protein and fat deposition. Nitrogen retention efficiency was higher at levels of 26.0 to 24.5% CP. It can be concluded that it is possible to reduce the PB content of feed for tambaqui fry from 32.0 to 24.5 per cent, and that reducing the PB content to levels between 27.5 and 24.5 per cent enhances fish performance, provided that the feed is properly supplemented with limiting essential amino acids.

Keywords: *Colossoma macropomum*, performance, protein nutrition, ideal protein.

SUMMARY

INTRODUCTION

The tambaqui (*Colossoma macropomum*) is one of the Brazilian fish with potential for zootechnical exploitation because, as well as having highly prized meat, it can reach a commercial size of around 1kg in the first year of rearing in captivity. Despite these characteristics, there is still a lack of information about their nutritional requirements, nutritional value and the effects of anti-nutritional factors in food (ingredients), with the aim of developing efficient, low-cost and environmentally friendly feeds (Hernandez, et al., 1995; Van Der Meer et al., 1995; Vidal Júnior et al., 1998; Silva et al., 2006; Pezzato et al., 2009; Cyrino et al., 2010).

Protein has been prioritised in studies on nutritional requirements for fish because, as well as being the most expensive component of feed, when levels exceed requirements they cause excessive nitrogen excretion and, together with phosphorus excretion, are the main causes of eutrophication in the aquatic environment (Sugiura, et al., 2001; Yamamoto et al., 2005; Bomfim et al., 2008).

For this species, there is a great difference in the recommended protein level in feed for the species in the fry stage, which has varied from 22 to 50 per cent of CP (Hernandez, et al., 1995; Van Der Meer et al., 1995; Vidal Júnior et al., 1998; Silva et al., 2006).

On the other hand, considering that fish metabolically require a quantitative balance between essential and non-essential amino acids, and that feeds formulated to meet dietary requirements based on crude protein usually contain excessive levels of non-limiting amino acids, the nitrogen content excreted into the environment has been high (Bomfim et al., 2008). In this sense, reducing the concentration of protein in the feed, with concomitant supplementation of limiting amino acids, based on the ideal protein concept (PARSONS; BAKER, 1994), would improve the balance between the amino acids in the feed, constituting a strategy for reducing the level of nitrogen excreted into the environment without jeopardising fish performance (Bomfim et al., 2008).

Research with Nile tilapia fry (Furuya et al., 2005; Bomfim et al., 2008) has shown that it is possible to reduce crude protein by up to four percentage points with the supplementation of limiting amino acids in industrial form, based on the ideal protein concept, with greater reductions resulting in worse performance. Araripe et al. (2011), on the other hand, did not observe compromised performance in tambatinga fry when they tested rations with reduced crude protein supplemented with industrial amino acids, which could be reduced by up to 12 percentage points. This variation in responses may indicate that the feasibility of reducing protein levels in the feed by supplementing with limiting amino acids may be species dependent.

One reason for this deterioration in performance is based on evidence that the excessive use of synthetic amino acids in feed may not guarantee the same efficiency of utilisation as those derived from the digestion of protein ingredients in the feed, due to their greater leaching capacity in the aquatic environment, higher stomach evacuation rate and rapid absorption compared to amino acids derived from protein in the feed, leading to a temporal imbalance of amino acids in the tissues specialised for protein synthesis (Zarate et al., 1999; Bomfim et al., 2008).

For tambaquis, there is also a lack of information on the feasibility of reducing the crude protein content of feed, with supplementation of limiting amino acids based on the ideal protein concept.

Considering the wide variation in the recommended protein content of rations for fry of this species (Pezzato et al., 2004; Vidal Júnior et al., 1998; Oishi et al., 2010), in order to determine how many percentage points the content of this nutrient can be reduced by supplementing amino acids, a preliminary trial should be carried out to determine the dietary level of crude protein that provides the best performance (control ration). This procedure would standardise the experimental conditions (temperature, feeding regime, etc.) between experiments, and would provide practical applicability of the results under different conditions in which the protein level that provides the best performance may differ from that determined under the experimental conditions in the reference studies.

CHAPTER 1

TAMBAQUI, PROTEIN AND AMINO ACIDS

1.1 Tambaqui General Characteristics

The tambaqui*, Colossoma macropomum* (Cuvier, 1818), is a fish native to the Amazon River basins, native to the Solimões, Madeira and Orinoco rivers (BALDISSEROTTO; GOMES, 2010). Belonging to the Characidae family and the Serrasalminae subfamily, it was the first Amazonian fish species to attract the attention of a large number of researchers and aquaculture producers due to its potential for fish farming (ARAÚJO-LIMA; GOULDING, 1998).

There are records indicating that the tambaqui is the second largest scale fish in South America, measuring up to 90 cm in length and weighing 30 kg (Gomes et al*.*, 2010; Lopera-Barrero et al., 2011), surpassed in size only by the Pirarucu (*Arapaima gigas*) (KUBITZA, 2004; FRACALOSSI; CYRINO, 2013). This species has shown high productivity in captivity, efficient growth potential, ease of obtaining fry and commercial value (VILLACORTA-CORREA, 1997; ARAÚJO-LIMA; GOULDING, 1998), an omnivorous/frugivorous/zooplankton eating habit (Claro-Júnior et al., 2004), the possibility of breeding in high densities and the ability to survive in environments with low oxygen concentrations (Melo et al., 2001).

In 1972, in north-eastern Brazil, precisely in Pentecostes/Ceará, tambaqui was introduced by the National Department of Works Against Droughts (DNOCS) in the ponds of the Ichthyological Research Centre (Silva et al., 1984). It is currently the most farmed native species in the country, with production in 24 of the 27 states, reaching around 111,000 tonnes in 2011 (MAPA, 2012). The North and Centre-West regions stand out as the biggest producers, while some states in the Northeast include Maranhão, Piauí, Sergipe and Bahia (Kubitza et al., 2012).

In Maranhão, official figures indicate a production of 1,603 tonnes in 2010. A recent survey carried out by Acqua Imagem found that just three large fish farmers in the Mearim Valley together produce around 410 tonnes of fish per year (Kubitza et al., 2012).

Many of the existing reference works group tambaqui together with pacu, pirapitinga and/or their hybrids as round fish, due to the similarities in the production system and the market for these species (Barros et al., 2011). The rearing of round fish species in polycultures has been carried out with the aim of minimising risks, since there are still no defined production technologies for each of these species or the best rearing conditions, as highlighted by Barros et al. (2011).

Currently, existing production practices are based on empirical experiences, linked to the monitoring of fish farmers, and little scientific information on the rearing of this species (Cavero et al., 2009; Barros et al., 2011). Complementary information has therefore been sought to improve the productive and economic performance of fish farming (Nunes et al., 2006; Chagas et al., 2007; Silva et al., 2007).

1.2 Protein and Amino Acids in Fish Nutrition

Protein is the main organic constituent of fish tissue, making up 65 to 75 per cent of total body dry matter (LALL;

ANDERSON, 2005; BORGHESI, 2008). It acts in the formation of specialised compounds such as enzymes and hormones, participates in the renewal and formation of protein tissue, and is one of the main sources of metabolic energy for fish (LOGATO, 2000; WILSON, 2002; MACHADO, 2004; LALL; ANDERSON, 2005; FRACALOSSI; CYRINO, 2013). In addition to these attributes, it is considered one of the most expensive items in feed formulation (BOSCOLO, 2003; Sá et al., 2006; BARROS, 2010).

For the growth, reproduction and other physiological functions of fish to occur normally, it is necessary that the nutritional requirements are adequately met, taking into account the differences between species (ROTTA, 2002). The foods selected by fish in the wild and under extensive production conditions (phytoplankton and zooplankton) have been of high quality and are the main source of food under these rearing conditions. In production conditions under more intensive regimes, where natural food is restricted, all essential nutrients must be obtained from the food provided, which must be highly digestible and in proportions that allow maximum utilisation (Jobling et al., 2001).

The different protein sources are not identical nutritionally and biologically. The biological value of a protein varies according to the composition of the amino acids and their respective availability (Anderson et al., 1995; Masumoto et al., 1996).

According to Sampaio (1998), if the diet contains a deficient amount of protein or an inadequate composition or bioavailability of amino acids, it can reduce growth, feed efficiency and fish weight due to the mobilisation of protein from some tissues to maintain vital functions. Excess protein in diets is used as a source of energy and can also reduce feed intake, increase nitrogen excretion into the water, which has a negative impact on the environment, and increase production costs.

Research carried out with tambaquis in the fry stage shows a wide variation in crude protein recommendations in diets, from 22 to 50 % CP (Vidal Júnior et al., 1998; Silva et al., 2006; Oishi et al., 2010; Santos et al., 2010), indicating the need for further research.

Considering that fish do not have a true protein requirement, but rather a quantitative balance of essential and non-essential amino acids, a diet formulated based on crude protein may not meet nutritional needs for all amino acids (CRAIG; HELFRICH, 2009; BARROS, 2010). Because of this, the concept of ideal protein has been advocated in the formulation of fish diets, which consists of an exact balance of amino acids in order to meet individual requirements for maintenance and production (CHUNG; BAKER, 1992; Ruchimat et al., 1997; NRC, 2011; FRACALOSSI; CYRINO, 2013).

In this concept, the essential amino acids are related to the reference amino acid, lysine, mainly because it is found in industrial form, there is a significant amount of information on its requirement, it is quick to analyse and it is used to a greater extent for the synthesis of body protein (VAN DER MEER; VERDEGEM, 1996; Abimorad et al., 2008). In this sense, based on the dietary requirement for lysine, the dietary requirement for the other essential amino acids is estimated (WILSON, 1989).

One nutritional strategy that has proven to be both productive and environmentally efficient is to reduce the crude protein content of feed by supplementing it with industrial amino acids, based on the ideal protein concept. This practice is based on reducing surplus levels of amino acids that are above the requirement associated with the supplementation of limiting amino acids in industrial form. As a result, there is a consequent improvement in the balance between amino acids, a reduction in the catabolism of non-limiting amino acids and in the excretion of nitrogen into the aquatic environment, without compromising productive performance (Bomfim et al., 2008; ARARIPE, 2009).

The feasibility of this practice was demonstrated in experiments with different fish species, as shown in Table 1. However, it should be noted that the extent to which it may be feasible to reduce protein levels in the feed, with supplementation of limiting amino acids, may be species and phase dependent. For Nile tilapia fry, the experiments carried out indicated that it is feasible to reduce the protein content by up to four and three percentage points for the fry and growth/termination phases, respectively. For rainbow trout and tambatinga there are indications that a reduction of ten to twelve percentage points is possible, respectively.

Table 1 - Reduction of crude protein in fish diets

Species	Live weight (g)	Protein reduction (%)	Supplemented amino acid*
Oreochromis niloticus [1]	0,8-16,32	32 -28	L/M/T
Oreochromis niloticus [2]	1,83- 12.46	32-28	L/M/T
Oreochromis niloticus [3]	4,4-127,26	30-27,5	L/M/T
Oreochromis niloticus [4]	34,63 273,90	27-24,3	L/M/T
Oreochromis niloticus [5]	105,7 493,05	26,7-24,5	L/M/T/A
P. brachypomus x *C. macropomum*[6]	6,521-65,079	40-28	L/M/T
Oncorhynchus mykiss [7]	25,20-272,9	45-35	L/M/T/G

[1]Bomfim et al. (2008);[2] Quadros et al. (2009);[3] Furuya et al. (2005);[4] Botaro et al. (2007);[5] Righetti et al. (2011);[6] Araripe et al. (2011);[7] Gaylor et al. (2009).

*Supplemented amino acids: L = L-lysine; M = DL-methionine; T = L-threonine; G = Glycine and A = L-arginine.

One factor that may influence the viability of this nutritional strategy is the implications of using industrial amino acids in diets, such as their greater leaching capacity in the aquatic environment, high stomach evacuation rate and rapid absorption compared to amino acids from protein sources in the feed, which in these cases result in an imbalance of amino acids at protein synthesis sites and, consequently, a reduction in the efficiency of protein utilisation (CHAKRABORTY, 1998; Righetti et al., 2011). To minimise this problem, it has been recommended to increase feeding frequency (TANTIKITTI; MARCH, 1995).

References

ABIMORAD, E. G.; SQUASSONI, G. H.; CARNEIRO, D. J. Apparent digestibility of protein, energy, and amino acids in some selected feed ingredients for pacu *Piaractus mesopotamicus*. **Aquaculture**, v.14, n.4, p.374 - 380, 2008.

ARARIPE, M. N. B. A.; ARARIPE, H. G. A.; LOPES, J. B.; CASTRO, P. L.; BRAGA, T. E. A.; FERREIRA, A. H. C.; ABREU, M. L. T. Reduction of crude protein with supplementation of amino acids in rations for tambatinga fry. **Revista Brasileira de Zootecnia**, v.40, n.09, p.1845-1850, 2011.

ARARIPE, M. N. B. A. **Reduction of crude protein and ratios of digestible methionine+cystine and threonine to digestible lysine in rations for tambatinga fry**. Thesis (Doctorate in Animal Science) - Federal University of Piauí, Teresina/PI, 2009.

ANDERSON, J. S.; LALL, S. P.; ANDERSON, D. M.; McNIVEN, M. A. Availability of amino acids from various fish meals fed to Atlantic salmon (*Salmo salar*). **Aquaculture**, v.138, p.291-301, 1995.

ARAÚJO-LIMA, C.; GOULDING, M. **The fruits of the tambaqui: ecology, conservation and cultivation in Amazonia.** Brasília, DF. Tefé Sociedade Civil Mamirauá/CNPq, 1998. 186 p.

BALDISSEROTTO, B.; GOMES, L. C. **Native species for fish farming in Brazil**. Santa Maria: UFSM, 2ª Ed, 2010. 175p.

BARROS, A. F. **Socio-economic and zootechnical analysis of fish farming in the micro-region of Baixada Cuiabana - MT.** 2010. 129f. Thesis (Doctorate in Aquaculture) UNESP Aquaculture Centre - CAUNESP, Jaboticabal, SP.

BARROS, A. F.; MARTINS, M. I. E. G.; SOUZA, O. M.. Characterisation of fish farming in the Baixada Cuiabana micro-region, Mato Grosso, Brazil. **Boletim do Instituto de Pesca**, São Paulo, v. 37, n. 3, p. 261 - 273, 2011.

BORGHESI, R. **Protein and energy requirements and biological value of feed for sea bream *Salminus brasiliensis***. Thesis (Doctorate in Agronomy) - University of São Paulo, Piracicaba/SP, 2008.

BOMFIM, M. A. D.; LANNA, E. A. T.; DONZELE, J. L.; ABREU, M. L. T.; RIBEIRO, F. B. QUADROS, M. Reduction of crude protein with amino acid supplementation, based on the ideal protein concept, in rations for nile tilapia fry. **Revista Brasileira de Zootecnia**, v.37, n.10, p.1713-1720, 2008.

BOSCOLO, W.R. **Waste flour from the tilapia filleting industry in the feeding of Nile tilapia (*Oreochromis niloticus L.*).** 2003. 98f. Thesis (Doctorate in Animal Production) - State University of Maringá UEM, Maringá, PR.

BOTARO, D.; FURUYA, W. M.; SILVA, L. C. R.; SANTOS, L. D.; SILVA, T. S. C.; SANTOS, V. G. Dietary protein reduction based on the ideal protein concept for nile tilapia (*Oreochromis niloticus*) raised in net pens. **Revista Brasileira de Zootecnia**, v.36, n.3, p.517-525, 2007.

CAVERO, B. A. S.; RUBIM, M. A. L.; PEREIRA, T. M. Commercial farming of tambaqui *Colossoma macropomum* (Cuvier, 1818) In: TAVARES- DIAS, M. (Org.). **Manejo e sanidade de peixes em cultivo**. Macapá: EMBRAPA Amapá, 2009. p. 33- 46.

CHAGAS, E. C. GOMES, L. C.; JUNIOR, H. M.; ROUBACH, R. Productivity of tambaqui raised in tank-nets with different feeding rates. **Ciência Rural**, Santa Maria, v. 37, n. 4, p. 1109-1115, 2007.

CHAKRABORTY, S. C.; CHAKRABORTY, S. Effect of dietary protein level on excretion of ammonia in Indian major carp (*Labeo rohita*), fingerlings. **Aquaculture Nutrition,** v. 4, p.47-51, 1998.

CHUNG, T. K.; BAKER, D. H. Ideal amino acid pattern for 10-kilogram pigs. **Journal of Animal Science**, v. 70, p.3102-3111, 1992.

CRAIG, S.; HELFRICH, L. A. Understanding Fish Nutrition, Feeds, and Feeding. **Virginia Cooperative Extension Service Publication,** v.1, p.420-456, 2009.

CYRINO, J. E. P.; BICUDO, A. J. A.; SADO, R. Y.; BORGHESI, R.;

DAIRIKI, J. K. Fish farming and the environment - the use of environmentally friendly feed in fish farming. **Revista Brasileira de Zootecnia**, v.39, p.68-87, 2010.

FRACALOSSI, D. M.; CYRINO, J. E. P. **Nutriaqua: Nutrition and feeding of species of interest to Brazilian aquaculture**. Florianópolis: Brazilian Society of Aquaculture and Aquatic Biology, 2013.

FURUYA, W. M.; BOTARO, D.; MACEDO, R. M. G; SANTOS, V. G.; SILVA, L. C. R.; SILVA T. C.; FURUYA, V. R. B.; SALES, P. J. P.. Application of the ideal protein concept to reduce protein levels in diets for Nile tilapia (*Oreochromis niloticus*). **Revista Brasileira de Zootecnia** v.34, n.5, p.1433-1441, 2005.

GAYLOR, T. G.; BARROWS, F. T. Multiple amino acid supplementations to reduce dietary protein in plantbased rainbow trout, *Oncorhynchus mykiss,* feeds. **Aquaculture**, v.287, p.180-184, 2009.

GOMES, L. C.; SIMÕES, L. N.; ARAÚJO-LIMA, C. A. R. M. 2010. Tambaqui (*Colossoma macropomum*), p.175-204. In: Baldisserotto, B.; Gomes, L.C. (Org.). **Native species for fish farming in Brazil**. 2ª . Ed. Editora UFSM, Santa Maria.

HERNANDEZ, M.; TAKEUCHI, T.; WATANABE, T. Effect of dietary energy sources on the utilisation of protein by *Colossoma macropomum* fingerlings. **Fisheries Science** 61 (3), p. 507-511, 1995.

JOBLING, M.; GOMES, E.; DIAS, J. Feed types, manufacture and ingredients. In: HOULIHAN, D.; BOUJARD, T.; JOBLING, M. (Ed.). **Food intake in fish.** Oxford: Blackwell Science, 2001. p. 25-48.

KUBITZA, F. Coletânea de informações aplicadas ao cultivo de tambaqui, do pacu e de outros peixes redondos. **Panorama da Aquicultura**, Rio de Janeiro, v.14, n.83, May/June 2004.

KUBITZA, F.; CAMPOS, J. L.; ONO, E. A.; ISTCHUK, P. I. Panorama of Fish Farming in Brazil: Statistics, species, production centres and factors limiting the expansion of the activity. **Panorama da Aquicultura**, Rio de Janeiro. vol. 22, n°132. July/August, 2012.

LALL, S. P.; ANDERSON, S. Amino acid nutrition of salmonids: Dietary requirements and bioavailablity. **Cahiers Options Méditerranéennes**, Montpellier, v. 63, p.73-90, 2005.

LOGATO, P.V.R. **Nutrição e alimentação de peixes de água doce.** Editora Aprenda Fácil: Viçosa, MG, 2000, 128p.

LOPERA-BARRERO, N. M.; RIBEIRO, R. P.; POVH, J. A.; VARGAS, L. D. M.; POVEDA-PARRA, A. R.; DIGMAYER, M. 2011. The main species produced in Brazil, 143-215. In: Lopera-Barrero, N.M.; Ribeiro, R.P.; Povh, J.A.; Vargas, L.D.M.; Poveda-Parra, A.R.; Digmayer, M. **Production of aquatic organisms: an overview in Brazil and worldwide**. Agrolivros, Guaíba.

MACHADO, C. C. **Protein requirement in the diet of gilthead seabream** *Salmius brasiliensis* **fry**. 2004. 44f. Dissertation (Master's Degree in Aquaculture) - Federal University of Santa Catarina - UFSC, Florianópolis, SC.

MASUMOTO, T.; RUCHIMAT, T.; ITO, Y.; HOSOKAWA, H.; SHIMENO, S. Amino acid availability values for several protein sources for yellowtail (*Seriola quinqueradiata*). **Aquaculture**, v.146, p.109-119, 1996.

MAPA. Ministry of Fisheries and Aquaculture. **Fisheries and Aquaculture Bulletin: Brazil 2012**. Brasília, 2012. 129 p.

MELO, L. A. S.; IZEL, A. C. U.; RODRIGUES, F. M. **Rearing tambaqui (*Colossoma macropomum*) in clay ponds/dams in the state of Amazonas.** Manaus, EMBRAPA, Western Amazonia, 2001, 25p.

NUNES, Z. M. P.; LAZZARO, X.; PERET, A. C. Influence of initial biomass on the growth and productivity of fish in a polyculture system. **Ciênc. Agrotec.**, Lavras, v. 30, n. 6, p. 1083-1090, 2006.

NATIONAL RESEARCH COUNCIL. **Nutrient requirements of fish**. Washington: National Academic Press, 2011.

OISHI, C. A.; NWANNA, L. C.; PEREIRA FILHO, M. Optimum dietary protein requirement for Amazonian Tambaqui, *Colossoma macropomum* Cuvier, 1818, fed fish meal free diets. **Acta Amazonica**, v.40, n.4, p.757-762, 2010.

PARSONS, C. M.; BAKER, D.H. The concept and use of ideal proteins in the feeding of nonruminants. In: REUNIÃO ANUAL DA SOCIEDADE BRASILEIRA DE ZOOTECNIA, 31, Maringá, 1994. **Proceedings**. Maringá: SBZ, 1994. p.119-128.

PEZZATO, L. E.; BARROS, M. M.; FRACALOSSI, D. M.; CYRINO, J. E. P. Fish nutrition. In: CYRINO, J.E.P.; URBINATTI, E.C.; FRACALOSSI, D.M.; CASTAGNOLLI, N. (Ed.). **Special topics in intensive tropical freshwater fish farming.** São Paulo: TecArt, 2004. ch.5, p.75-169.

PEZZATO, L. E.; BARROS, M. M.; FURUYA, W. M. Nutritional value of foods used in the formulation of rations for tropical fish. **Revista Brasileira de Zootecnia**, v.38, p.43-51, 2009 (special supplement).

QUADROS, M.; LANNA, E. A. T.; DONZELES, J. L.; ABREU, M. L. T.; RIBEIRO, F. B.; TAKISHITA, S. S. Crude protein reduction and digestible methionine+cystine and threonine to digestible lysine ratios in diets for Nile tilapia fingerlings. . **Revista Brasileira de Zootecnia,** v.38, n.8, p.1400-1406, 2009.

RIGHETTI, J. S.; FURUYA, W. M.; CONEJERO, C. I.; GRACIANO, T. S.; VIDAL, L. V. O.; MICHELLATO, M. Protein reduction in diets for nile tilapia through amino acid supplementation based on the ideal protein concept. **Revista Brasileira Zootecnia**, v.40, n.03, p.469-476, 2011.

ROTTA, M. A. 2002. **Energy and protein utilisation by fish**, Corumbá: Embrapa Pantanal, 24p.

RUCHIMAT, T.; MASUMOTO, T.; HOSOKAWA, H.; ITOH, Y.; SHIMENO, S. Quantitative lysine requirement of yellowtail (*Seriola quinqueradiata*), **Aquaculture,** v.158, p.331- 339, 1997.

SAMPAIO, A. M. B. M. **Energy: protein ratio in the nutrition of tucunaré *(Cichla sp)***. 1998. 49p. Master's thesis - University of São Paulo, Piracicaba.

SILVA, L. C. R; FURUYA, W. M.; SANTOS, L. D.; SANTOS, V. G.; SILVA, T. S. C.; PINSETTA, P. J. Levels of threonine in feed for Nile tilapia, **Revista Brasileira de Zootecnia**, v. 35, n. 4, p. 12581264, 2006.

SILVA, A. M. D.; GOMES, L. C.; ROUBACH, R. Growth, yield, water and effluent quality in ponds with different management during tambaqui juvenile production. **Pesquisa Agropecuária Brasileira**, Brasília, v. 42, n. 5, p. 733-740, 2007.

SILVA, J. W. B., NOBRE, M. I. S., PINHEIRO, F. A. E SOBRINHO, A. C. 1984. Results of a polyculture experiment of Tambaqui Colossoma macropomum, Cuvier, 1818, tilapia hybrid, (*Oreochromis hornorum* TREW x *O. niloticus* L., 1766) and mirror *carp Cyprinus carpio* L., 1758 vs Speculares. **DNOCS Technical Bulletin**, Fortaleza, 42 (1): 63-89.

SUGIURA, S. H.; GABAUDAN, J.; DONG, F. M.; HARDY, R. W. Dietary microbialphytase supplementation and utilisation of phosphorus, trace minerals and protein bu rainbow trout *Oncorhynchus mykiss* (Walbaum) fed soybean meal-based diets. **Aquaculture Research.** v.32, p.583-592, 2001.

TACON, A. G. J. **Nutritional fish pathology**. Rome: FAO, 1992. p.1- 75.

TANTIKITTI, C.; MARCH, B. E. Dynamics of plasma free amino acids in rainbow trout (Oncorhynchus mykiss) under variety of dietary conditions. Fish Physiology Biochemistry, v.14, p.179-194, 1995.

VAN DER MEER, M. B.; MACHIELS, M. A. M.; VERDEGEM, M. C. J.The effect of dietary protein level on growth, protein utilisation and

body composition of *Colossoma macropomum* (Cuvier). **Aquaculture Research** n.26, p. 901-909, 1995.

VAN DER MEER, M. B.; VERDEGEM, M.C.J. Comparison of amino acid profiles of feeds and fish as a quick method for selection of feed ingredients: a case study for *Colossoma macropomum* (Cuvier) at different dietary fish meal/soya meal ratios. **Aquaculture Research**, Oxford, v. 27, p.531-538, 1996.

VILLACORTA-CORREA, M. A. **Study of the age and growth of the tambaqui *Colossoma macropomum* (Characiformes: Characidae) in Central Amazonia, by analysing seasonal marks on mineralised structures and microstructures on otoliths**. 1997, 214p. Dissertation (Master's Degree) - National Institute for Amazonian Research, Manaus.

WILSON, R. P. Amino acids and Proteins. In: HALVER, J.E.; HARDY, R.W. **Fish Nutrition**. 3rd ed.San Diego: Elsevier Academic Press, 2002. p. 143-179.

WILSON, R. P. Amino acids and proteins. In: HALVER, J.H. (Ed.). **Fish nutrition.** San Diego: Academic Press. 1989. p.112-153.

YAMAMOTO, T.; SUGITA, T.; FURUITA, H. Essential amino acid supplementation to fish meal-based diets with low protein to energy ratios improves the protein utilisation in juvenile raimbow trout *Oncorhinchus mykiss*. **Aquaculture** v.246, n.1-4, p.379-391, 2005.

ZARATE, D. D.; LOVELL, R.T; PAYNE, M. Effects of feeding frequency and rate of stomach evacuation on utilisation of dietary free and protein-bound lysine for growth by channel catfish *Ictalurus punctatus*. **Aquaculture Nutrition**, v.5, p.17-22, 1999.

CHAPTER 2

CRUDE PROTEIN IN FEED FOR TAMBAQUI JUVENILES

Summary

The aim was to evaluate the effects of crude protein (CP) levels in feed for tambaqui fry. 750 fry weighing 0.35 ± 0.09 g were used in a completely randomised design with 6 treatments, 5 replicates and 25 fish per experimental unit. The treatments consisted of six isoenergetic, isocaloric and isophosphoric diets, with 20; 24; 28; 32; 36 and 40 per cent CP. Performance parameters, body composition, body protein and fat deposition and nitrogen retention efficiency were evaluated. There was no effect of CP levels on feed intake and carcass yield. CP intake increased linearly and the protein efficiency ratio decreased linearly. Weight gain, feed conversion and specific growth rate improved up to the estimated levels of 31.57, 28.90 and 31.12 per cent CP, respectively. Body moisture was not influenced by BW levels. There was a quadratic effect of crude protein levels on body fat, body fat deposition, body protein, body protein deposition and nitrogen retention efficiency, with a minimum estimated level for the first two of 26.55; 23.77; and a maximum estimated level for the others of 29.26; 32.50; 27.21% of CP, respectively. It was concluded that the recommended CP level for tambaqui fry is 31.57%, which corresponds to a digestible energy (DE):CP ratio of 9.50 kcal DE/g CP, as it provides better weight gain.

Keywords: *Colossoma macropomum,* performance, feed management, protein nutrition.

2.1 Introduction

The tambaqui (*Colossoma macropomum*) is a native species that has aroused interest in fish farming due to its desirable characteristics such as:

adaptation to captivity, high prolificacy, rapid growth, omnivorous eating habits, acceptance of artificial food, white-coloured meat with few intramuscular spines and acceptance by the consumer market (Mendonça et al., 2009).

Despite their productive characteristics, there is still a lack of information on nutritional requirements (Dairiki & Silva, 2011). As far as dietary crude protein content is concerned, there is a wide divergence in the recommended protein level for the fry stage, which has ranged from 22 to 50 per cent CP (Vidal Júnior et al., 1998; Silva et al., 2006; Oishi et al., 2010; Santos et al., 2010).

Diets deficient in protein, with inadequate amino acid composition or bioavailability, can lead to a reduction in growth rate and feed efficiency due to the mobilisation of protein from some tissues to maintain vital functions. Excess protein in diets is used as a source of energy and can also reduce feed intake, increase nitrogen excretion in water and increase production costs (Fracalossi & Cyrino, 2013). Therefore, further research is needed to accurately determine protein levels in feed for each fish species and rearing stage.

The aim of this study was to evaluate the effects of crude protein levels in diets for tambaqui fry.

2.2 Material and Methods

The experiment lasted 45 days, from September to October 2012, in the Fish Nutrition Research Sector of the Centre for Agricultural and Environmental Sciences of the Federal University of Maranhão (UFMA), located in the municipality of Chapadinha, Maranhão.

A total of 750 tambaqui fry (*Colossoma macropomum*) with an initial weight of 0.35 ± 0.09 g were used, distributed in a completely randomised design with six treatments, five replications and twenty-five fish per experimental unit.

The treatments consisted of six isoenergetic, isocaloric and isophosphoric experimental diets containing different levels of crude protein - CP (20; 24; 28;

32; 36 and 40.0 %) (Table 1). The minimum level of amino acids, calculated as a percentage of the crude protein content of the feed, was estimated based on the nutritional requirements suggested for nile tilapia by Furuya et al. (2010).

Table 1. Percentage and chemical composition of the rations in experiment 1 (natural matter)

Ingredients (%)	Crude Protein Level (%)					
	20,00	24,00	28,00	32,00	36,00	40,00
Soya meal (45%)	33,863	44,648	55,323	65,997	76,672	87,347
Maize	57,580	46,440	35,376	24,313	13,249	2,186
Soya oil	3,848	4,247	4,625	5,003	5,381	5,759
L-Lysine-HCl (78.4%)	0,048	0,000	0,000	0,000	0,000	0,000
DL-Methionine (99%)	0,060	0,098	0,138	0,178	0,217	0,257
L-Threonine (98.5%)	0,093	0,118	0,145	0,173	0,200	0,227
Calcitic limestone	0,298	0,236	0,177	0,118	0,059	0,000
Bicalcium phosphate	3,140	3,142	3,145	3,148	3,151	3,154
Vitamin and Mineral Supplements (1)	0,100	0,100	0,200	0,300	0,400	0,500
Vitamin C (2)	0,050	0,050	0,050	0,050	0,050	0,050
Salt	0,500	0,500	0,500	0,500	0,500	0,500
Antioxidant (BHT)	0,020	0,020	0,020	0,020	0,020	0,020
Calculated composition (3)						
Crude Protein (%)	20,00	24,00	28,00	32,00	36,00	40,00
Digestible Protein (%) (4)	17,97	21,74	25,51	29,28	33,05	36,82
Digestible Energy (kcal/kg) (4)	3000,0 0	3000,00	3000,00	3000,00	3000,00	3000,00
Ether extract (%)	6,48	6,65	6,80	6,96	7,11	7,26
Crude fibre (%)	2,79	3,17	3,54	3,92	4,29	4,67
Total Ca (%)	0,98	0,98	0,98	0,98	0,98	0,98
P disp (%) 4	0,70	0,70	0,70	0,70	0,70	0,70
Na Total (%)	0,22	0,22	0,22	0,22	0,22	0,22
Lysine (%) Total	1,115	1,352	1,625	1,897	2,170	2,442
Lysine (%) Digest(4)	1,043	1,264	1,520	1,776	2,032	2,288
Met + Cist (%) Total	0,683	0,822	0,961	1,101	1,240	1,380
Met + Cyst (%) Digest. (4)	0,626	0,757	0,888	1,019	1,151	1,282
Total threonine (%)	0,879	1,060	1,241	1,422	1,604	1,785
Threonine Digest (%) (4)	0,803	0,971	1,140	1,308	1,477	1,645
Total tryptophan (%)	0,248	0,309	0,370	0,430	0,491	0,552
Tryptophan Digest (%) (4)	0,229	0,286	0,342	0,398	0,455	0,511
ED/PB ratio (Kcal/g)	15,00	12,50	10,71	9,37	8,33	7,50

(1)Commercial vitamin and mineral supplement (5 kg/t), with guarantee levels per kilogram of product: Vit. A, 1,200,000 IU; Vit. D3, 200,000 IU; Vit. E, 1,200 mg; Vit. K3, 2,400 mg; Vit. B1, 4,800 mg; Vit. B2, 4,800 mg; Vit. B6, 4,800 mg; Vit. B12, 4,800 mg; Vit. C, 48 g; folic acid, 1,200 mg; Ca pantothenate, 12,000 mg; Vit. C, 48,000 mg; biotin, 48 mg; choline chloride, 108 g; niacin, 24,000 mg; Fe, 50,000 mg; Cu, 3,000 mg; Mn, 20,000 mg; Zn, 30,000 mg; I, 100 mg; Co, 10 mg; Se, 100 mg. (2)Vit. C: calcium 2-monophosphate salt of ascorbic acid, 42% active ingredient. (3)Based on the values proposed by Rostagno et al. (2011). (4)Based on the phosphorus availability coefficients proposed for Nile tilapia by Furuya et al. (2010); and those for corn, soya meal and soya oil proposed by Vidal Júnior (2000).

The fry were kept in 30 polyethylene aquariums, with a volumetric capacity of 500 litres and a useful volume of 300 litres, equipped with individual water supply and bottom drainage systems, arranged in a recirculation system.

The water used to supply the aquariums came from an artesian well at the Centre for Agricultural and Environmental Sciences at the Federal University of Maranhão.

The water temperature was monitored daily at 7.30am and 5.30pm using a mercury bulb thermometer graduated from 0 to 50°C. The pH, dissolved oxygen content in the water and toxic ammonia were monitored every seven days using a pH meter, an oximeter and a commercial toxic ammonia test kit, respectively.

The ingredients in the diets were pelletised using a meat grinder, ground manually and sieved, leaving pellets 3 to 4 mm in diameter (Fracalossi & Cyrino, 2013).

The diets were fed *ad libitum* daily, in five meals (08:00; 10:30; 13:00; 15:30 and 18:00). At each meal, the diets were fed in small quantities, with successive instalments, allowing maximum intake until apparent satiety, thus avoiding over- or under-supply.

The aquariums were cleaned twice a day to remove faeces by siphoning, after water temperature readings.

The following performance and body composition indices were evaluated: feed consumption (FC), crude protein consumption (CPB), protein efficiency ratio (PER), weight gain (WG), feed conversion (FC), specific growth rate (SGR), carcass yield (CR), body composition (moisture, protein and fat content), body protein and fat deposition rates, and nitrogen retention efficiency.

CR was calculated from the amount of feed consumed from the beginning to the end of the experiment. CPB was obtained by multiplying feed consumption by the protein percentage, divided by 100. TEP was obtained by dividing weight gain by crude protein intake.

WG was obtained from the difference in weight between the average final and initial weights. AC was calculated by dividing apparent feed consumption by weight gain. The equation below was used to determine EER, using logarithmic transformations:

$$TCE = \frac{\text{LN of final weight (g) - LN of initial weight (g) x 100}}{\text{experiment time (days)}}$$

For the body analyses, a sample of 50 fish were stunned, euthanised and frozen at the beginning, and all the fish from each experimental unit at the end of the experiment.

Before freezing, six fish per experimental unit, with weights corresponding to the average weight of the respective unit, were eviscerated to determine carcass yield. The carcass yield (%) was calculated by dividing the weight of the eviscerated fish by the weight of the fish with viscera, multiplied by 100.

After freezing, the fish (carcasses and viscera) were freeze-dried, ground in a ball mill and packed in containers for laboratory analysis.

The carcass samples were analysed for their body chemical composition (moisture, protein and body fat content) according to the procedures described by Silva & Queiroz (2003).

Body protein and fat deposits were calculated as the difference between the final and initial body protein or fat, respectively, in mg, divided by the experimental period in days.

The nitrogen retention efficiency, expressed as a percentage, was calculated as the difference between the final and initial body nitrogen, divided by the total nitrogen consumed, multiplied by 100.

The performance and body composition variables were subjected to the normality (Cramer-Von Mises) and homoscedasticity (Brown-Forsythe) tests. The data was subjected to analysis of variance at the five per cent probability level.

The effects of crude protein levels were analysed using polynomial regression models of 1ª and 2ª order or discontinuous "Linear Response Plateau" (LRP), depending on the best fit obtained for each variable.

Statistical analyses were carried out using SAS 9.0 *software* (2002).

2.3 Results and Discussion

The values obtained during the experimental period of 26.8 ± 0.48°C and 28.6 ± 0.65°C for water temperature in the morning and afternoon, respectively, 5.8 ± 0.25 for pH, 6.7 ± 0.83 mg/L for dissolved oxygen and 0.40 ± 0.12 mg/L for total ammonia, remained within the recommended range for breeding the species (temperature between 28°C. pH between 4 and 6 mg/L; dissolved oxygen above 3 mg/L; concentrations of 0.46 mg/L); pH between 4 and 6 mg/L; dissolved oxygen above 3 mg/L; concentrations of 0.46 mg/L of total ammonia), according to Gomes et al. (2010) and Mendonça et al. (2012).

It was observed that varying the crude protein (CP) content of the feed had no influence (P>0.05) on the feed consumption and carcass yield of the fish. However, the other variables were influenced (P<0.05) by the treatments (Table 2).

Table 2. Performance, feed efficiency and carcass yield of tambaqui fry as a function of the protein content of the feed.

Variables	Crude protein level (%)						CV	p>f[1] [2]
	20	24	28	32	36	40		
Feed consumption (g)	14,90	13,90	13,58	15,39	14,20	13,47	9,56	0,2584
Crude protein intake (mg)[3]	2,98	3,33	3,80	4,92	5,11	5,38	11,07	<.0001
Protein efficiency ratio[4]	3,43	3,41	3,56	2,77	3,02	2,82	11,92	0,0131
Weight gain (g)[5,6]	10,23	11,38	13,51	13,66	15,45	15,17	14,82	0,0023
Feed conversion (g/g)[7,8]	1,46	1,25	1,02	1,13	0,92	0,89	11,62	<.0001
Specific growth rate (%/day)[9,10]	7,59	7,83	8,16	8,24	8,45	8,40	3,86	0,0016
Carcass yield (%)	83,19	83,28	83,06	85,95	85,20	85,42	2,60	0,1541

(1)Coefficient of variation (%). (2)Significance of the "F" test from the analysis of variance. (3)Linear effect: Y = 0.131X + 0.297 (r^2 = 0.615). [(4)]Linear effect: Y = - 0.035X + 4.241 (r^2 = 0.949). [(5)]Linear Response Plateau effect: Y = 14.84 - 0.4105X (R^2 = 0.916); plateau: Y = 14.84. [(6)]Linear effect: Y = 0.2647x + 5.2919 (r^2 = 5.2919). [(7)]Linear Response Plateau effect: Y = 0.9736 - 0.055X (R^2 = 0.873); plateau: Y = 0.974. [(8)]Linear effect: Y = - 0.0266x + 1.911(r^2 = 0.8455). [(9)]Linear Response Plateau effect: Y = 8.377 - 0.723X (R^2 = 0.958); plateau: Y = 8.377. [(10)]Linear effect: Y = 0.0428x + 6.8281 (r^2 = 0.9026).

In the case of feed consumption, the probable reason for the lack of variation in this variable is due to the fact that the diets were isoenergetic, since the energy level of the diet can limit fish consumption (Carneiro, 1990). Vidal Júnior et al. (1998), evaluating the protein requirements of tambaqui in the 30 to 250 gram phase, also found no differences in feed consumption with increasing protein levels in diets for this species.

Since feed intake did not vary, crude protein intake increased (P<0.01) linearly with the increase in protein levels in the diet. The protein efficiency ratio (PER) decreased linearly (P<0.05) as the protein levels in the diet increased.

Weight gain (WG) increased (P<0.01) linearly as the levels of PB in the feed increased. Despite the linear variation, the "Linear Response Plateau" (LRP) model was the best fit to the data, estimating the level of CP at 31.57%, from which a plateau of 14.84 g occurred (Table 1 and Figure 1).

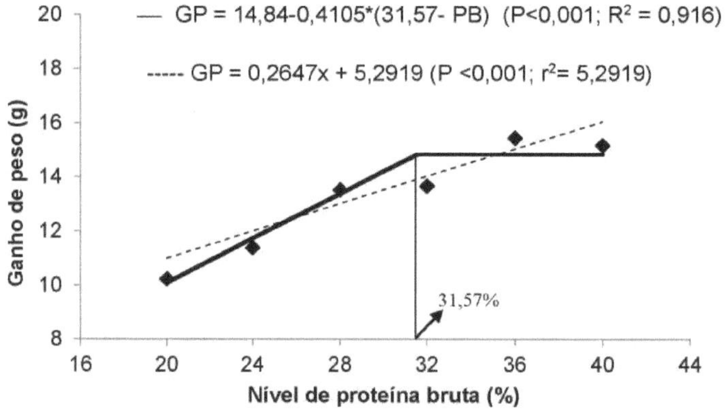

Figure 1. Graphical representation of the weight gain of tambaqui fry as a function of the protein level in the feed.

Similar effects were also observed by Sá & Fracalossi (2002) and Bomfim et al. (2005), who found a linear increase in the weight gain of piracanjuba (*Brycon orbignyanus)* and curimbatá (*Prochilodus affins*) fry, respectively, with a plateau occurring from one of the intermediate levels of CP assessed.

On the other hand, these results differ from those obtained by Vidal Júnior et al. (1998) and Furuya et al. (2000) with tambaquis in the 30 to 250 gram phase and Nile tilapia fry, respectively, who found a quadratic effect on weight gain with variations in the level of protein in the diet, indicating that excessive levels of protein can be detrimental to growth, as the animals may require energy to metabolise and excrete the excess protein.

In the case under study, considering that there was no variation in feed consumption and, consequently, in energy and other nutrients (isoenergetic, isocaloric and isophosphoric diets), it can be inferred that weight gain was influenced only by the variation in crude protein consumption. In this sense, the consumption of crude protein provided by diets with levels above 31.57%, despite not providing significant improvements in weight gain, did not have a negative influence on feed consumption, as the protein was probably used for energy purposes, corroborating the results observed for TEP.

Although the variation in TEP was linear, it can be seen that, numerically, the reduction in this variable became more evident when diets containing 28% CP were used, indicating a reduction in the efficiency of protein utilisation for body protein deposition and an increase in the proportion directed towards energy, gluconeogenesis or lipogenesis (Salhi et al., 2004; Fracalossi & Cyrino, 2013).

Similar results have also been observed in other fish studies in which an increase in TEP was observed as the protein concentration in the diet was reduced (Bicudo et al., 2010; Costa et al., 2009; Cotan et al., 2006; Vidal Júnior et al., 1998).

On the other hand, these results differ from those observed by Sá & Fracalossi, (2002) with piracanjuba fry, who found no significant differences in TEP as a result of varying the concentration of protein in the diet.

Different observations were also made by Borghesi (2008), in an experiment with juvenile sea bream, and by Bomfim et al. (2005), in curimbatá fry, who found that up to a certain protein level there was an improvement in TEP, suggesting that the rate of protein efficiency rises until the protein requirement is reached by the fish, with a reduction in its value at subsequent protein levels.

Similar to what was observed for weight gain, the specific growth rate also increased ($P<0.01$) linearly ($Y = 0.0428x + 6.8281$; $P <0.001$; $r^2 = 0.9026$) as the levels of PB in the feed increased. Despite the linear effect, the "Linear Response Plateau" (LRP) model was the one that best fitted the data, estimating at 31.12% the level of CP from which there was a plateau at 8.377 %/day.

These results are similar to those obtained by Souza et al. (2011) with juvenile peacock bass, in which they observed that the EER showed an increasing linear effect, indicating better results as the protein level of the diet increased. Similarly, Vásquez-Torres et al. (2011) observed that juvenile pirapitinga (*Piaractus brachypomus*) showed the highest specific growth rates (SGR) when fed a diet containing high levels of crude protein (up to 32% CP).

In the case of feed conversion, despite the linear effect ($Y = - 0.0266x + 1.911$; $P <0.001$; $r^2 = 0.8455$), the "Linear Response Plateau" - LRP model was the best fit for the data, estimating the level of CP at 28.91 per cent, from which there was a plateau at 0.974 g/g.

In a study conducted by Bittencourt et al. (2010) with piavuçu fry and Souza et al. (2011) with juvenile sea bass, apparent feed conversion improved linearly as the concentration of protein in the diet increased.

However, in some studies, feed conversion varied quadratically as a function of the variation in protein levels, such as those carried out by Sá &

Fracalossi (2002) with piracanjuba and by Bomfim et al. (2005) with curimbatá. In others, the variation in protein levels did not affect the fish's feed conversion, as in the experiments carried out by Botaro et al. (2007) with Nile tilapia, Vidal Júnior et al. (1998) with tambaqui and Signor et al., (2004) with jundiá.

The better values observed for feed conversion in fish fed higher levels of CP may be attributed to their higher growth rates (Bomfim et al., 2005), especially in lean tissue, since the fish's feed consumption was similar between the protein levels evaluated. This hypothesis could be based on the dry matter content of the protein tissue, which is lower than that of the fat tissue, thus incorporating a greater amount of water into the carcass (Fracalossi & Cyrino, 2013).

Protein levels had no influence ($P>0.05$) on the carcass yield (CR) of tambaqui fry. The CR values obtained in this experiment were between 83.06 and 85.95, similar to those reported by Reidel et al. (2010) (CR between 78 and 87%) in jundiá and Signor et al. (2010) (CR between 84.69 and 85.01%) in pacus. However, the CR results are higher than those reported by Melo et al. (2003), who obtained values of between 80.0 and 82.5 per cent in jundiá; but lower than those observed by Reidel et al. (2004) for male and female curimbatá (averages of 91.26 and 89.40 per cent, respectively).

Body moisture content ($P>0.05$) was not influenced by the variation in CP content (Table 3). On the other hand, Bomfim et al. (2005) working with curimbatá fry and Cotan et al. (2006) with tambiú lambari fry observed that fish fed diets consisting of low levels of crude protein had lower body moisture contents compared to treatments with higher levels of CP.

There was a quadratic effect ($P<0.01$) of CP levels on body fat content, which decreased up to the estimated level of 26.55% CP, with a minimum value of 2.36% (Table 3). These results differ from those observed by Bomfim et al. (2005) and Fernandes et al. (2001), who found a reduction in body fat as CP levels increased in curimbatá and pacu, respectively.

With regard to body protein content, there was a quadratic effect ($P<0.05$) of crude protein levels on this variable, which improved up to the estimated level of 29.26 per cent CP, with a maximum value of 14.26 per cent.

Bittencourt et al. (2010), in their research with piavuçu (*Leporinus macrocephalus*) fry, found higher protein percentages in the carcasses of fish fed 30% in relation to treatments with higher levels of CP. Fernandes et al. (2000) observed that an increase in crude protein levels influenced an increase in body protein levels in pacu and tambaqui fry, respectively.

There was a quadratic effect ($P<0.01$) of CP levels on body fat deposition, which decreased up to the estimated level of 23.77% CP, with a minimum value of 6.01% (Table 3). These results corroborate work carried out by Fernandes et l. (2000) with pacu, and by Sá & Fracalossi, (2002) with piracanjuba, indicating that when tambaqui fry were fed lower protein levels, they made use of the additional energy for fat deposition.

Body protein deposition showed a quadratic effect ($P<0.05$) with the reduction in CP, which improved up to the estimated level of 32.50 per cent CP, with a maximum value of 44.29 per cent (Table 3). These results corroborate the observations of Oishi et al. (2010) and Santos et al. (2010) that protein deposition rates are closely related to the concentration of protein in the diet.

There was a quadratic effect ($P<0.01$) of crude protein levels on nitrogen retention efficiency - NRE, which improved up to the estimated level of 27.21 per cent CP, with a maximum value of 44.79 per cent. It can be seen that the reduction in nitrogen retention efficiency observed with the use of lower and higher levels of CP than the estimated maximum point is probably related, in the first case, to the increase in the proportion of protein used for maintenance needs in relation to that used for protein deposition, and, in the second case, due to the use of the additional protein consumed for energy purposes.

A similar quadratic effect on ERN as a function of dietary protein variation was also observed by Bomfim et al. (2005) with curimbatás. With tilapia, Furuya et al. (2005) found that this variable decreased linearly as the protein content of the diet increased. On the other hand, Fernandes et al. (2001) observed that variations in dietary crude protein levels did not affect the ERN of pacus, and Cotan et al. (2006) found a higher ERN in fish fed the lowest protein level in tambiú lambari fry.

2.4 Conclusion

The crude protein recommendation for tambaqui fry is 31.57%, which corresponds to a digestible energy (DE):CP ratio of 9.50 kcal DE/g CP, as it provides better weight gain.

Thank you

To the Foundation for Research Support and Scientific and Technological Development of Maranhão, for financial support.

References

BICUDO, A. J.A.; SADO, R.Y.; CYRINO, J.E.P. Growth performance and body composition of pacu *Piaractus mesopotamicus* (Holmberg 1887) in response to dietary protein and energy levels. **Aquaculture Nutrition**, v.16, n.2, p.213-222, 2010.

BITTENCOURT, F.; FEIDEN, A.; SIGNOR, A.A.; BOSCOLO, W.R.; FREITAS, J.M.A. Protein and energy in rations for piauçu fry. **Revista Brasileira de Zootecnia**, v.39, n.12, p.2553-2559, 2010.

BOMFIM, M.A.D.; LANNA, E.A.T.; SERAFINI, M.A.; RIBEIRO, F.B.; PENA, K.S. Crude protein and digestible energy in diets for curimbatá (*Prochilodus affins*) fry. **Revista Brasileira de Zootecnia**, v.34, n.6, p.1795-1806, 2005.

BORGHESI, R. **Protein and energy requirements and biological value of feed for sea bream *Salminus brasiliensis*.** 2008. 95p. Thesis (Doctorate) - University of São Paulo, Piracicaba, 2008.

BOTARO, D.; FURUYA, W. M.; SILVA, L. C. R.; SANTOS, L. D.; SILVA, T. S. C.; SANTOS, V. G. Dietary protein reduction based on the ideal protein concept for Nile tilapia (Oreochromis nilotucus) raised in net pens. **Revista Brasileira de Zootecnia,** v.36, n.3, p.517-525, 2007.

CARNEIRO, D.J. Effect of temperature on protein and energy requirements in diets for pacu fry, *Piaractus mesopotamicus* (HOLMBERG, 1887). São Carlos, SP: UFSCAR, 1990. 55p. Thesis (Doctorate in Ecology) - Federal University of São Carlos, 1990.

COTAN, J. L. V.; LANNA, E. A. T.; BOMFIM, M. A. D.; DONZELES, J. L.; RIBEIRO, F. B.; SERAFINI, M. A. Levels of digestible energy and crude protein in rations for tambiú lambari fry. **Revista Brasileira de Zootecnia,** v.35, n.3, p.634-640, 2006.

COSTA, M. L. S.; MELO, F. P.; CORREIA, E. S. Effects of different protein levels in feed on growth in Nile Tilapia (*Oreochromis niloticus* Linnaeus, 1757), Chitralada variety, raised in net ponds. **Boletim do Instituto de Pesca**, São Paulo, 35(2): 285 - 294, 2009.

DAIRIKI, J. K.; SILVA, T. B. A. **Literature review: nutritional requirements of tambaqui - compilation of works, formulation of suitable feed and future challenges.** Manaus: Embrapa Amazônia Ocidental, 2011. 44p (Embrapa Amazônia Ocidental, 91).

FERNANDES, J. B. K.; CARNEIRO, D. J.; SAKOMURA, N. K. Sources and Levels of Crude Protein in Diets for Pacu Fry (*Piaractus mesopotamicus* **Revista Brasileira de Zootecnia**, 29(3):646-653, 2000.

FERNANDES, J. B. K.; CARNEIRO, D. J.; SAKOMURA, N. K. Sources and

Levels of Crude Protein in Diets for Juvenile Pacu (*Piaractus mesopotamicus*). **Revista Brasileira de Zootecnia**, 30(3):617-626, 2001.

FRACALOSSI, D. M.; CYRINO, J. E. P. **Nutriaqua: Nutrition and feeding of species of interest to Brazilian aquaculture**. Florianópolis: Brazilian Society of Aquaculture and Aquatic Biology, 2013.

FURUYA, W. M.; BOTARO, D.; MACEDO, R. M. G.; SANTOS, V. G.; SILVA, L. C. R.; SILVA, T. C.; FURUYA, V. R. B.; SALES, P. J. P..

Application of the ideal protein concept to reduce protein levels in diets for Nile tilapia. **Revista Brasileira de Zootecnia**, v.34, n.5, p.1433-1441, 2005.

FURUYA, W. M.; HAYASHI, C.; FURUYA, V. R. B.; SOARES, C. M. Protein requirements for Nile tilapia (*Oreochromis niloticus*) fry. **Revista Brasileira de Zootecnia**, v.29, p.1912-1917, 2000.

FURUYA, W. M.; PEZZATO, L. E.; BARROS, M. M.; BOSCOLO, W. R.; CYRINO, J. E. P.; FURUYA, V. R. B.; FEIDEN, A. **Brazilian tables for tilapia nutrition**. Toledo: GFM, 2010. 100p.

GOMES, L.C.; SIMÕES, L.N.; ARAÚJO-LIMA, C.A.R.M. Tambaqui (*Colossoma macropomum*). In: Baldisserotto, B.; Gomes, L.C. (Org.). **Native species for fish farming in Brazil.** 2ª . Ed. Editora UFSM, Santa Maria. 2010. p.175-204.

MELO, J.F.B.; BOIJINK, C.L.; RADUNZ NETO, J. Efeito da alimentação na composição química da carcaça do jundiá *Rhamdia quelen*. **Revista Biodiversidade Pampeana**, v.1, n.1, p.12-23, 2003.

MENDONÇA, P. P. COSTA, P. C. POLESE, M. F. VIDAL JÚNIOR, M. V.; ANDRADE, D. R. Effect of phytase supplementation in the feeding of juvenile tambaqui (*Colossoma macropomum*). **Archivos de Zootecnia**, 61 (235): 437-448. 2012.

MENDONÇA, P. P.; FERREIRA, R. A; VIDAL JUNIOR, M. V.; ANDRADE, D. R.; M. V. B. SANTOS, M. V. B; FERREIRA, A. V.; REZENDE, F. P. 2009. Influence of photoperiod on the development of juvenile tambaqui (*Colossoma macropomum*). **Archivos de Zootecnia**, 58 (223): 323-331, 2009.

OISHI, C.A.; NWANNA, L.C.; PEREIRA FILHO, M. Optimum dietary protein requirement for Amazonian Tambaqui, *Colossoma macropomum* Cuvier, 1818, fed fish meal free diets. **Acta Amazonica**, v.40, n.4, p.757-762, 2010.

REIDEL, A.; ROMAGOSA, E.; FEIDEN, A.; BOSCOLO, W. R.; COLDEBELLA, A.; SIGNOR, A. A. Body yield and chemical composition of jundi fed different levels of protein and energy in the diet, reared in net pens. **Revista Brasileira de Zootecnia**, v.39, n.2, p.233-240, 2010.

REIDEL, A.; OLIVEIRA, L. G.; PIANA, P. A.; LEMAINSKI, D.; BOMBARDELLI, R.A.; BOSCOLO, W. R. Evaluation of yield and morphometric characteristics of male and female curimbatá *Prochilodus lineatus* (VALENCIENNES, 1836) and piavuçu *Leporinus macrochephalus* (GARAVELLO & BRITSKI, 1988). **Revista Varia Scientia**, v.4, n.8, p.71-78, 2004.

FRACALOSSI, D. M.; CYRINO, J. E. P. **Nutriaqua: Nutrition and feeding of species of interest to Brazilian aquaculture**. Florianópolis: Brazilian Society of Aquaculture and Aquatic Biology, 2013.

ROSTAGNO, R. S.; ALBINO, L. F. T.; DONZELE, J. L.; GOMES, P. G.; OLIVEIRA, R. F.; LOPES, D. C.; FERREIRA, A. S.; BARRETO, S. L. T.; EUCLIDES, R. F. **Brazilian tables for poultry and pigs - food composition and nutritional requirements**. 2. ed. Viçosa: Federal University of Viçosa, 2011.

SALHI, M.; BESSONART, M.; CHEDIAK, G.; BELLAGAMBA, M.; CARNEVIA, D. Growth, feed utilisation and body composition of black catfish, Rhamdia quelen, fry fed diets containing different protein and energy levels. **Aquaculture**, v. 231, p. 435-444, 2004.

SANTOS, L.; PEREIRA FILHO, M.; SOBREIRA, C.; ITUASSÚ, D.; FONSECA, F. A. L. Protein requirement of juvenile tambaqui (*Colossoma macropomum)* after food deprivation. **Acta Amazonica**, v. 40(3), 597-604, 2010.

SÁ, M. V. C.; FRACALOSSI, D. M. Protein requirement and energy/protein ratio for piracanjuba (*Brycon orbignyanus*) fry. **Revista Brasileira de Zootecnia**, v.31, n.1, p.1-10, 2002.

SIGNOR, A.; SIGNOR, A. A.; FEIDEN, A.; BOSCOLO, W. R.; REIDEL, A.; HAYASHI, C. Crude protein requirement of jundiá (*Rhamdia quelen*) fry. **Varia Cientia**, v.4, n.1, p.79-89, 2004.

SIGNOR, A. A.; BOSCOLO, W. R.; FEIDEN, A.; BITTENCOURT, F.; COLDEBELLA, A.; REIDEL, A. Protein and energy in pacu feed raised in net pens. **Revista Brasileira de Zootecnia**, v.39, n.11, p.2336-2341, 2010.

SILVA, L. C. R; FURUYA, W. M.; SANTOS, L. D.; SANTOS, V. G.; SILVA, T. S. C.; PINSETTA, P. J. Levels of threonine in feed for Nile tilapia, **Revista Brasileira de Zootecnia,** v. 35, n. 4, p. 12581264, 2006.

SOUZA, J. H.; FRACALOSSI, D. M.; GARCIA, A. S.; RIBEIRO, F. F; TSUZUKI, M. Y. Zootechnical and economic performance of juvenile sea bass fed diets containing different protein concentrations. **Pesquisa Agropecuária Brasileira**, Brasília, v.46, n.2, p.190- 195, feb. 2011.

VÁSQUEZ-TORRES, W.; PEREIRA FILHO, M.; ARIAS-CASTELLANOS, J.A. Optimum dietary crude protein requirement for juvenile cachama *Piaractus brachypomus.* **Ciência Rural**, v.41, n.12, p.2138-2189, 2011.

VIDAL JÚNIOR, M.V.; DONZELE, J. L.; CAMARGO, A. C. S.; ANDRADE, D. R.; SANTOS, L. C. Protein levels for Tambaqui (*Colossoma *macropomum* Cuvier, 1818) from 30 to 250 g live weight. Tambaqui performance. **Revista Brasileira de Zootecnia**, V. 27, n.3, p. 421-426, 1998.

VIDAL JÚNIOR, M. V. **Techniques for determining the digestibility of nutrients in food for tambaqui (*Colossoma macropomum*)**. 2000. 96p. Thesis (Doctorate). Federal University of Viçosa, Viçosa.

CHAPTER 3

REDUCTION OF CRUDE PROTEIN, WITH AMINO ACID SUPPLEMENTATION, IN FEED FOR TAMBAQUI FRY

Summary

The aim was to evaluate the effects of reducing crude protein (CP) with amino acid supplementation, based on the ideal protein concept, in feed for tambaqui fry. 750 fry with an initial weight of 0.44 ± 0.14 g were used in a completely randomised design with 6 treatments, 5 replications and 25 fish per experimental unit. The treatments consisted of 6 isoenergetic, isocaloric, isophosphoric and isolysin rations with 32.0; 30.5; 29.0; 27.5; 26.0; 24.5% CP. Performance variables, body composition, body protein and fat deposition and nitrogen retention efficiency were evaluated. As the CP content decreased from 27.5 per cent CP, there was an improvement in weight gain, specific growth rate, feed conversion and protein efficiency rate. Feed consumption was higher in fish fed diets containing 26% CP, followed by 29.0 and 27.5% CP, while the other levels did not vary. CP consumption decreased with the use of lower CP diets and carcass yield was not influenced by the treatments. Reducing the protein content of the feed did not influence body composition, protein and fat deposition. Nitrogen retention efficiency was higher at levels of 26.0 to 24.5% CP. It can be concluded that it is possible to reduce the CP content of feed for tambaqui fry from 32.0 to 24.5 per cent, and that the reduction in CP content does not affect body composition, protein deposition and body fat.

PB levels of between 27.5 and 24.5 per cent boost fish performance, as long as the feed is properly supplemented with limiting essential amino acids.

Keywords: amino acids, *Colossoma macropomum,* performance, protein nutrition, fish farming.

3.1 Introduction

Proteins have been prioritised in studies on nutritional requirements for fish because, as well as being the most expensive component of feed, when they are higher than required they cause excessive nitrogen excretion and, together with phosphorus, are the main factors responsible for eutrophication in the aquatic environment (Yamamoto et al., 2005; Bomfim et al., 2008).

On the other hand, considering that fish metabolically require a quantitative balance between essential and non-essential amino acids, and that feeds formulated to meet dietary requirements based on crude protein usually contain excessive levels of non-limiting amino acids (Bomfim et al., 2008), reducing the concentration of protein in feed, with concomitant supplementation of amino acids, based on the ideal protein concept would improve the balance between the amino acids in the diet, constituting a strategy for reducing the level of nitrogen excreted into the environment without harming the development and performance of the fish (Fracalossi & Cyrino, 2013).

Studies with Nile tilapia fry (Furuya et al., 2005; Bomfim et al., 2008) show that it is possible to reduce crude protein by up to four percentage points by supplementing limiting amino acids in industrial form, based on the ideal protein concept, with greater reductions resulting in worse performance. With tambatinga, Araripe et al. (2011) did not observe compromised performance in fry when they tested rations with reduced crude protein supplemented with industrial amino acids. This variation in responses may indicate that the feasibility of reducing protein levels in the feed, with supplementation of limiting amino acids, is species-dependent.

One reason for the worsening in performance with high reductions in crude protein content is based on evidence that the excessive use of industrial amino acids in feed may not guarantee the same efficiency of use as those derived from the digestion of protein ingredients in the feed, due to their greater leaching capacity, higher rate of stomach evacuation and rapid absorption compared to amino acids derived from feed protein, leading to a temporal imbalance of amino acids in tissues specialised for protein synthesis (Bomfim et al., 2008).

For tambaquis, there is a lack of information on the feasibility of reducing the crude protein content of feed, with supplementation of limiting amino acids based on the ideal protein concept.

The aim of this study was to assess the effects of reducing crude protein and supplementing limiting amino acids, based on the ideal protein concept, in feed for tambaqui (*Colossoma macropomum*) fry.

3.2 Material and Methods

The experiment lasted 45 days, from December 2012 to February 2013, in the Fish Nutrition Research Sector of the Centre for Agricultural and Environmental Sciences of the Federal University of Maranhão (UFMA), located in the municipality of Chapadinha, Maranhão.

A total of 750 tambaqui fry (*Colossoma macropomum*) with an initial weight of 0.44 ± 0.14 g were used, distributed in a completely randomised design consisting of six treatments, five replications and twenty-five fish per experimental unit.

The treatments consisted of a control diet with 32 per cent CP (recommendation obtained in the first experiment) and five more experimental diets with a gradual reduction of one and a half percentage points in the crude protein content compared to the reference diet (30.5; 29.0; 27.5; 26.0 and 24.5 per cent), all isoenergetic, isocaloric and isophosphoric and isolysinic (Table 1).

Table 1. Percentage and chemical composition of the rations in experiment 2 (natural matter)

Ingredients (%)	Crude Protein Level (%)					
	32,00	30,50	29,00	27,50	26,00	24,50
Soya meal (45%)	66,641	62,629	58,617	54,605	50,593	46,581

Maize	23,664	27,652	31,641	35,629	39,618	43,606
Soya oil	5,029	4,794	4,558	4,323	4,088	3,852
L-Lysine-HCl (78.4%)	0,000	0,123	0,247	0,370	0,493	0,617
DL-Methionine (99%)	0,172	0,208	0,244	0,280	0,315	0,351
L-Threonine (98.5%)	0,164	0,220	0,276	0,332	0,388	0,443
L-Tryptophan (99%)	0,000	0,022	0,043	0,065	0,087	0,109
Calcitic limestone	0,110	0,133	0,155	0,177	0,199	0,222
Bicalcium phosphate	3,148	3,149	3,149	3,149	3,149	3,149
Vitamin and Mineral Supplement[1]	0,500	0,500	0,500	0,500	0,500	0,500
Vitamin C[2]	0,050	0,050	0,050	0,050	0,050	0,050
Salt	0,500	0,500	0,500	0,500	0,500	0,500
Antioxidant (BHT)	0,020	0,020	0,020	0,020	0,020	0,020
Calculated composition[3]						
Crude Protein (%)	32,00	30,50	29,00	27,50	26,00	24,50
Digestible Protein (%)[4]	29,27	27,86	26,45	25,04	23,63	22,21
Digestible Energy (kcal/kg)[4]	3000,00	3000,00	3000,00	3000,00	3000,00	3000,00
Ether extract (%)	6,97	6,81	6,66	6,50	6,35	6,19
Crude fibre (%)	3,94	3,80	3,65	3,51	3,37	3,22
Total Ca (%)	0,98	0,98	0,98	0,98	0,98	0,98
P disp (%)[2]	0,70	0,70	0,70	0,70	0,70	0,70
Na Total (%)	0,22	0.22	0.22	0.22	0.22	0,22
Lysine (%) Total	1,914	1,908	1,902	1,896	1,889	1,883
Lysine (%) Digest[4]	1,791	1,791	1,791	1,791	1,791	1,791
Met + Cist (%) Total	1,102	1,099	1,096	1,093	1,091	1,088
Met + Cyst (%) Digest.[4]	1,020	1,020	1,020	1,020	1,020	1,020
Total threonine (%)	1,424	1,420	1,416	1,413	1,409	1,405
Threonine Digest (%)[4]	1,309	1,309	1,309	1,309	1,309	1,309
Total tryptophan (%)	0,434	0,433	0,431	0,430	0,428	0,427
Tryptophan Digest (%)[4]	0,402	0,402	0,402	0,402	0,402	0,402
ED/PB ratio (Kcal/g)	0,638	0,636	0,634	0,632	0,630	0,628
Ratio Met+Cist /Lysine Dig.	56,9	56,9	56,9	56,9	56,9	56,9
Threonine/Dig. Lysine Ratio	73,1	73,1	73,1	73,1	73,1	73,1
Tryptophan/Dig. Lysine Ratio	22,4	22,4	22,4	22,4	22,4	22,4

(1)Commercial vitamin and mineral supplement (5 kg/t), with guarantee levels per kilogram of product: Vit. A, 1,200,000 IU; Vit. D_3, 200,000 IU; Vit. E, 1,200 mg; Vit. K_3, 2,400 mg; Vit. B_1, 4,800 mg; Vit. B_2, 4,800 mg; Vit. B_6, 4,800 mg; Vit. B_{12}, 4,800 mg; Vit. C, 48 g; folic acid, 1,200 mg; Ca pantothenate, 12,000 mg; Vit. C, 48,000 mg; biotin, 48 mg; choline chloride, 108 g; niacin, 24,000 mg; Fe, 50,000 mg; Cu, 3,000 mg; Mn, 20,000 mg; Zn, 30,000 mg; I, 100 mg; Co, 10 mg; Se, 100 mg. (2)Vit. C: calcium 2-monophosphate salt of ascorbic acid, 42% active ingredient. (3)Based on the values proposed by Rostagno et al. (2011). (4)Based on the phosphorus availability coefficients proposed for Nile tilapia by Furuya et al. (2010); and those for corn, soya meal and soya oil proposed by Vidal Júnior (2000).

In order to make the isolysine feeds digestible, industrial lysine (L- lysine-HCl 78.4%) was added as the protein level was reduced. The other amino acids were supplemented as their ratios to digestible lysine fell below those estimated in the ideal protein based on the requirement values for nile tilapia proposed by Furuya et al. (2010).

The fry were kept in 30 polyethylene aquariums, with a volumetric capacity of 500 litres and a useful volume of 300 litres, equipped with individual water supply and bottom drainage systems, arranged in a recirculation system.

The water used to supply the aquariums came from an artesian well at the Centre for Agricultural and Environmental Sciences at the Federal University of Maranhão.

The water temperature was monitored daily at 7.30am and 5.30pm using a mercury bulb thermometer graduated from 0 to 50°C. The pH, dissolved oxygen content in the water and toxic ammonia were monitored every seven days using a pH meter, an oximeter and a commercial toxic ammonia test kit, respectively.

The ingredients in the diets were pelletised using a meat grinder, ground manually and sieved, leaving pellets 3 to 4 mm in diameter (Fracalossi & Cyrino, 2013).

The diets were fed *ad libitum* daily, in five meals (08:00; 10:30; 13:00; 15:30 and 18:00). At each meal, the diets were fed in small quantities, with successive instalments, allowing maximum intake until apparent satiety, thus avoiding over- or under-supply.

The aquariums were cleaned twice a day to remove faeces by siphoning, after water temperature readings.

The following performance and body composition indices were evaluated: feed consumption (FC), crude protein consumption (CPB), protein efficiency ratio (PER), weight gain (WG), feed conversion (FC), specific growth rate (SGR), carcass yield (CR), body composition (moisture, protein and fat content), body protein and fat deposition rates, and nitrogen retention efficiency.

CR was calculated from the amount of feed consumed from the beginning to the end of the experiment. CPB was obtained by multiplying feed consumption by the protein percentage, divided by 100. TEP was obtained by dividing weight gain by crude protein intake.

WG was obtained from the difference in weight between the average final and initial weights. AC was calculated by dividing apparent feed consumption by weight gain. The equation below was used to determine EER, using logarithmic transformations:

$$TCE = \frac{\text{LN of final weight (g) - LN of initial weight (g) x 100}}{\text{experiment time (days)}}$$

For the body analyses, a sample of 50 fish were stunned, euthanised and frozen at the beginning, and all the fish from each experimental unit at the end of the experiment.

Before freezing, six fish per experimental unit, with weights corresponding to the average weight of the respective unit, were eviscerated to determine carcass yield. The carcass yield (%) was calculated by dividing the weight of the eviscerated fish by the weight of the fish with viscera, multiplied by 100.

After freezing, the fish (carcasses and viscera) were freeze-dried, ground in a ball mill and packed in containers for laboratory analysis.

The carcass samples were analysed for their body chemical composition (moisture, protein and body fat content) according to the procedures described by Silva & Queiroz (2003).

Body protein and fat deposits were calculated as the difference between the final and initial body protein or fat, respectively, in mg, divided by the experimental period in days.

The nitrogen retention efficiency, expressed as a percentage, was calculated as the difference between the final and initial body nitrogen, divided by the total nitrogen consumed, multiplied by 100.

The performance and body composition variables were subjected to the normality (Cramer-Von Mises) and homoscedasticity (Brown-Forsythe) tests. When these assumptions were met, the variables were subjected to analysis of variance and then compared using Duncan's test, considering a significance level of 5% or less.

For the variables of body fat deposition, body protein deposition and nitrogen retention efficiency, the requirements of normality and homoscedasticity were met after the logarithmic transformation Log (x).

Statistical analyses were carried out using SAS 9.0 *software* (2002).

3.3 Results and Discussion

The values obtained during the experimental period of 27.0 ± 0.40°C and 28.9 ± 0.55°C for water temperature in the morning and afternoon, respectively, 6.0 ± 0.30 for pH, 6.6 ± 0.36 mg/L for dissolved oxygen and 0.41 ± 0.12 mg/L for total ammonia, remained within the recommended range for breeding the species (temperature between 28°C. pH between 4 and 6 mg/L; dissolved oxygen above 3 mg/L; concentrations of 0.46 mg/L); pH between 4 and 6 mg/L; dissolved oxygen above 3 mg/L; concentrations of 0.46 mg/L of total ammonia), according to Gomes et al. (2010) and Mendonça et al. (2012).

In terms of feed consumption, only the fish fed diets containing 26.0 per cent CP were higher (P<0.05) than those fed the control diet. Fish fed diets containing 27.5 and 29.0 per cent CP had intermediate intakes (Table 2).

Table 2. Performance, feed efficiency and carcass yield of tambaqui fry as a function of reducing the protein content of the feed supplemented with amino acids.

Variables	Crude protein level (%)						. $CV^{(1)}$	$P>F_{(2)}$
	32,0	30,5	29,0	27,5	26	24,5		
CR	$30,24^{b}$	$29,67^{b}$	$32,90^{ab}$	$34,25^{ab}$	$37,22^{a}$	$28,80^{b}$	13,03	0,0955
CPB	$9,68^{a}$	$9,05^{a}$	$9,54^{a}$	$9,42^{a}$	$9,68^{a}$	$7,06^{b}$	12,38	0,0291
TEP	$1,36^{b}$	$1,86^{b}$	$1,70^{b}$	$3,49^{a}$	$4,04^{a}$	$4,30^{a}$	25,83	<0,0001
GP	$13,37^{c}$	$17,15^{bc}$	$17,25^{bc}$	$33,12^{ab}$	$39,67^{a}$	$30,69^{ab}$	39,75	0,0105
CA	$2,60^{a}$	$2,02^{ab}$	$2,57^{a}$	$1,06^{b}$	$0,96^{b}$	$0,96^{b}$	38,74	0,0024
TCE	$7,93^{b}$	$8,44^{b}$	$8,20^{b}$	$10,04^{a}$	$10,36^{a}$	$9,83^{a}$	9,66	0,0032
RC	83,23	85,40	83,36	82,83	83,78	84,83	3,87	0,7922

CR = Feed intake (g). CPB = Crude protein intake (mg). TEP = Protein efficiency ratio. WG = Weight gain (g). FC = Feed conversion (g/g). SCT = Specific growth rate (%/day). CR = Carcass yield (%). (1)Coefficient of variation (%). (2)Significance of the "F" test from the analysis of variance. (3)Averages followed by equal letters on the same line do not differ according to the Duncan test (P>0.05).

CP consumption varied (P<0.05) with the reduction in CP content, with fish fed diets containing 24.5 per cent CP showing lower CP consumption compared to those fed the other experimental diets, which did not differ from each other (Table 2). Araripe et al. (2011) and Bomfim et al. (2008), on the other hand, found that the BW intake of monkfish and Nile tilapia, respectively, decreased as the protein content in diets supplemented with limiting amino acids decreased.

The protein efficiency ratio showed the best values (P<0.05) in the fish fed diets with levels between 27.5 and 24.5 per cent of CP compared to the other treatments, which did not differ from each other.

The increase in protein efficiency with the reduction in crude protein is due to the fact that the control diet (32% CP) is made up of excessive levels of protein-bound amino acids in relation to nutritional requirements, compared to rations with reduced CP supplemented with amino acids (Bomfim et al., 2008). These results are in line with the theory that fish, as observed in poultry and pigs, do not require protein, but rather an adequate balance of amino acids to meet maintenance and production needs (Ahmed and Khan, 2004; Furuya et al., 2004; Furuya et al., 2005; Orlando et al., 2005; Yamamoto et al., 2005).

In this study, it was observed that the tambaqui fry used industrial amino acids efficiently, although they are subject to low utilisation efficiency due to the possibility of leaching and rapid absorption (Baldisseroto, 2010). As the feed was provided in five meals a day and with successive feedings, this may have reconciled the differences between the absorption rates of amino acids from protein and industrial sources (Rodehutscord et al., 1997).

Tantikitti and March (1995) observed in rainbow trout that the higher the feeding frequency, the better the compatibility between the concentrations of industrial amino acids and those linked to protein sources, thus providing better protein efficiency.

In a study carried out by Araripe et al. (2011) with monkfish, reducing the PB content linearly improved the protein efficiency rate. On the other hand, Rigethitt et al. (2011), working with Nile tilapia, found that reducing protein levels had a quadratic effect on the protein efficiency ratio, so that the best estimated value for digestible protein was 24.25%.

On the other hand, Botaro et al. (2007) and Berge et al. (2002) found no effect of reducing the levels of protein supplemented with limiting amino acids on the protein efficiency rate of Nile tilapia and Atlantic salmon, respectively.

For weight gain (WG) and specific growth rate (SGR), the reduction in the CP content of the diets supplemented with amino acids to levels between 27.5 and 24.5 per cent showed higher values (P<0.05) than those fed the control ration (32.0 per cent CP). The other levels showed intermediate values and did not differ from each other (Table 2).

Table 3. Body composition, body protein and fat deposits and nitrogen retention efficiency of tambaqui fry fed diets containing different protein contents.

Variables	Crude protein level (%)							. CV(1)	P>F(2)
	Home	32,0	30,5	29,0	27,5	26	24,0		
UC	72,48	81,19	81,17	81,41	80,94	81,94	81,84	1,61	0,8026
GC	1,21	1,31	2,01	1,78	1,27	1,88	1,27	32,86	0,1164
PC	20,87	11,73	12,95	12,29	11,68	13,87	12,48	16,45	0,5580
DGC(3)	---	6,31	10,26	7,65	9,50	11,04	7,65	30,81	0,2635
CPS(3)	---	47,98	74,26	67,45	74,33	105,59	74,69	17,28	0,4591
ERN(3)	---	21,65[b]	32,07[ab]	29,56[ab]	35,44[ab]	47,88[a]	47,72[a]	18,16	0,1462

BA = Body Moisture (%). BF = Body fat (%). CP = Body protein (%). DGC = Body fat deposition (mg/day). CPD = Body protein deposition (mg/day). NRE = Nitrogen retention efficiency (%). (1)Coefficient of variation (%). (2)Significance of the "F" test from the analysis of variance. (3)Analyses were carried out using transformed data (log (x)). (4)Averages followed by equal letters on the same line do not differ according to the Duncan test (P>0.05).

The values obtained for TCE in this study were higher than the 3.67%/day observed by Aride et al. (2006), who assessed the growth and survival of juvenile tambaqui for 45 days; and those obtained by Silva et al. (2007), of 5.07%/day, with juvenile tambaqui for 45 days in net pens. It is possible that the values observed in this study were high because they were referring to growth in the initial phase, which is normally higher in the first month of rearing.

With regard to weight gain, this improvement may be associated with the higher feed consumption (energy and nutrients) observed at the respective protein levels, probably due to the better amino acid balance with the reduction in the PB content supplemented with limiting amino acids (Furuya et al., 2004)

Compared to experiments with other species, there is a similarity to those observed with grass carp (*Ctenopharyngodon idella)* by Gan et al. (2012). Kim (1997) in an experiment with rainbow trout, evaluating diets containing 35, 30, 25, 20, 15 and 10% CP, observed a linear increase (P<0.05) in the weight gain of the fish up to the 25% CP level supplemented with industrial arginine and methionine.

In other studies with similar objectives, however, no significant differences were observed in weight gain for tambatinga (Araripe et al., 2011) and Nile tilapia (Furuya et al., 2005; Bomfim et al., 2008; Botaro et al, 2007; Righetti et al., 2011; Quadros et al., 2009). Although no improvement in weight gain was observed, these results indicate that reducing the recommended protein content for the species is feasible, as long as adequate amino acid supplementation is provided.

With regard to feed conversion, the reduction in the CP content of the amino acid-supplemented rations to levels between 27.5 and 24.5 per cent CP showed better values (P<0.05) than those fed the other levels and the reference ration (32 per cent CP), which did not differ from each other (Table 2).

The variation in feed conversion in this study is similar to that obtained by Berge et al. (2002) with Atlantic salmon, in which they observed an improvement in the feed conversion of fish fed a diet supplemented with industrial amino acids.

However, Araripe et al. (2011) and Botaro et al. (2007) observed no significant differences in the feed conversion of tambatinga and Nile tilapia fry, respectively, with the reduction in protein content supplemented with limiting amino acids. On the other hand, Righetti et al. (2011) observed that reducing protein levels had a quadratic effect on the feed conversion of Nile tilapia.

The improvement in feed conversion at 27.5 to 24.5% CP levels can be attributed to the greater proportion of nutrients being used for production processes than for maintenance (Guillaume, et al., 2004; Fracalossi & Cyrino, 2013), as well as the efficient use of industrial amino acid levels in diets with lower protein levels (Furuya et al., 2004).

It was found that reducing the protein content of the feed did not influence (P>0.05) carcass yield (Table 2), body moisture (BH), body fat (BF), body protein (BP), body fat deposition (BFD) and body protein deposition (BDP). However, they did influence (P<0.05) the nitrogen retention efficiency (NRE) in tambaqui fry (Table 3).

These results differ from those observed by Yamamoto et al. (2005) with trout, in which they noted that rations with higher levels of protein showed lower levels of body fat, the result of which was associated with the caloric expenditure required by the animals to metabolise the excess amount of amino acids, resulting in lower net energy and a greater caloric increase (Fracalossi and Cyrino, 2013).

It was expected that there would be an improvement in the variables of body fat deposition and body protein deposition, as they showed statistical differences for weight gain as a function of the variation in the protein content of the diets. This may have been due to the value of the corresponding coefficient of variation (CV%) that these variables obtained, since body fat and body protein content, respectively, are taken into account when calculating DGC and DPC, as well as final weight.

With the reduction in the protein content of the feed, there was an improvement in the nitrogen retention efficiency (P<0.05) of the tambaqui fry at the lower protein levels, especially those of 26.0 to 24.5 per cent CP, compared to those fed with feed containing 32.0 per cent CP.

Similar responses have been observed in other studies with Nile tilapia (Furuya et al., 2005; Quadros et al., 2009) and rainbow trout (Yamamoto et al.,

2005), increasing protein retention due to better amino acid balance, and consequently reducing nitrogen excretion into the environment.

On the basis of these results, a nutritional strategy based on the preparation of rations with a reduced protein fraction, based on the ideal protein concept, is an efficient way of reducing the polluting potential of the activity without compromising fish performance (Furuya et al., 2005; Orlando et al., 2005; Yamamoto et al., 2005; Bomfim et al., 2008; Araripe et al., 2011).

3.4 Conclusion

It is possible to reduce the crude protein content of tambaqui fry feed from 32.0 to 24.5 per cent, as long as the feed is supplemented with limiting amino acids.

Reducing the CP content of feed supplemented with limiting amino acids to 24.5 per cent boosts fish performance.

Thank you

To the Foundation for Research Support and Scientific and Technological Development of Maranhão, for financial support.

References

AHMED, I.; KHAN, M. A. Dietary lysine requirement of fingerling Indian major carp, Cirrhinus mrigala(Hamilton). **Aquaculture**, v.235, p.499-511, 2004.

ARIDE, P. H. R.; ROUBACH, R.; NOZAWA, S. R.; VAL, A. L. Tambaqui growth and survival when exposed to different photoperiods. **Acta Amazônica**, v.36, n.3, p.381- 384, 2006.

ARARIPE, M. N. B. A.; ARARIPE, H. G. A.; LOPES, J. B.; CASTRO, P. L.; BRAGA, T. E. A.; FERREIRA, A. H. C.; ABREU, M. L. T. Reduction of crude protein with supplementation of amino acids in rations for tambatinga fry. **Revista Brasileira de Zootecnia**, v.40, n.09, p.1845-1850, 2011.

BALDISSEROTO, B. **Fisiologia de peixes aplicada à piscicultura**. 2.ed.

Santa Maria: UFSM, 2010.

BERGE, G.E.; SUEIER, H.; LIED, E. Effects of feeding Atlantic salmon (*Salmo salar*) imbalanced levels of lysine and arginine. **Aquaculture Nutrition**, v.8, p.239-248, 2002.

BOMFIM, M. A. D.; LANNA, E. A. T.; DONZELE, J. L.; ABREU, M. L. T.; RIBEIRO, F. B. QUADROS, M. Reduction of crude protein with amino acid supplementation, based on the ideal protein concept, in rations for nile tilapia fry. **Revista Brasileira de Zootecnia**, v.37, n.10, p.1713-1720, 2008.

BOTARO, D.; FURUYA, W. M.; SILVA, L. C. R.; SANTOS, L. D.; SILVA, T. S. C.; SANTOS, V. G. Dietary protein reduction based on the ideal protein concept for nile tilapia (*Oreochromis niloticus*) raised in net pens. **Revista Brasileira de Zootecnia**, v.36, n.3, p.517-525, 2007.

FRACALOSSI, D. M.; CYRINO, J. E. P. **Nutriaqua: Nutrition and feeding of species of interest to Brazilian aquaculture**. Florianópolis: Brazilian Society of Aquaculture and Aquatic Biology, 2013.

FURUYA, W. M.; BOTARO, D.; MACEDO, R. M G; SANTOS, V. G.; SILVA, L. C. R.; SILVA T. C.; FURUYA, V. R. B.; SALES, P. J. P. Application of the ideal protein concept to reduce protein levels in diets for Nile tilapia (*Oreochromis niloticus*). **Revista Brasileira de Zootecnia** v.34, n.5, p.1433-1441, 2005.

FURUYA, W. M.; PEZZATO, L. E.; BARROS, M. M.; BOSCOLO, W. R.; CYRINO, J. E. P.; FURUYA, V. R. B.; FEIDEN, A. **Brazilian tables for tilapia nutrition**. Toledo: GFM, 2010. 100p.

FURUYA, W. M.; PEZZATO, L. E.; BARROS, M. M.; PEZZATO, A. C.; FURUYA, V. R. B.; Miranda, E. C. Use of ideal protein concept for precision formulation of amino acid levels in fish-meal-free diets for juvenile Nile tilapia (*Oreochromis niloticus L.*). **Aquaculture Research**, v.35, p.1110-1116, 2004.

GAN, L; LIU, Y. J.; TIAN, L. X.; YANG, H. J.; YUE, Y. R.; CHEN, Y. J.; LIANG, J. J.; LIANG, G. Y. Effect of dietary protein reduction with lysine and methionnine supplementation on growth performance, body composition and total ammonia nitrogen excretion of juvenile grass carp, *Ctenopharyngodon idella.* **Aquaculture Nutrition,** v. 18, p. 589598, 2012.

GOMES, L.C.; SIMÕES, L.N.; ARAÚJO-LIMA, C.A.R.M. Tambaqui (*Colossoma macropomum*). In: Baldisserotto, B.; Gomes, L.C. (Org.). **Native species for fish farming in Brazil.** 2ª . Ed. Editora UFSM, Santa Maria. 2010. p.175-204.

GUILLAUME, J.; KAUSHIK, S.; BERGOT, P.; MÉTAILLER, R. **Nutrición y alimentación de peces y crustáceos**. Madrid: Mundi-Prensa, 2004. 475p.

KIM, K-IL. 1997. Re-evaluation of protein and amino acid requirements of rainbow trout (*Oncorhynchus mykiss*). **Aquaculture**, v.151, p. 3-7.

MENDONÇA, P. P. COSTA, P. C. POLESE, M. F. VIDAL JÚNIOR, M. V.; ANDRADE, D. R. 2012. Effect of phytase supplementation in the feeding of juvenile tambaqui (*Colossoma macropomum*). **Archivos de Zootecnia**, v. 61, p. 437-448.

ORLANDO, U. A. D.; OLIVEIRA, R. F. M.; DONZELE, J. L. FERREIRA, A. S.; SILVA, F. C. O.; GENEROSO, R. A. R.; VAZ, R. G. M. V.; SIQUEIRA, J. C. Crude protein levels and amino acid supplementation in rations for suckling pigs kept in a comfort environment

from 30 to 60kg. **Revista Brasileira de Zootecnia**, v.34, n.1, p.134-141, 2005.

QUADROS, M.; LANNA, E. A. T.; DONZELES, J. L.; ABREU, M. L. T.; RIBEIRO, F. B.; TAKISHITA, S. S. Crude protein reduction and digestible methionine+cystine and threonine to digestible lysine ratios in diets for Nile tilapia fingerlings. . **Revista Brasileira de Zootecnia,** v.38, n.8, p.1400-1406,

2009.

RIGHETTI, J. S.; FURUYA, W. M.; CONEJERO, C. I.; GRACIANO, T. S.; VIDAL, L. V. O.; MICHELLATO, M. Protein reduction in diets for nile tilapia through amino acid supplementation based on the ideal protein concept. **Revista Brasileira Zootecnia**, v.40, n.03, p.469-476, 2011.

RODEHUTSCORD, M.; BECKER, A.; PACOTE, M.; PFEFFER, E. Response of rainbow trout (*Oncorhinchus mykiss*) to supplements of individual essential amino acids in a semipurified diet, including an estimation of the maintenance requirement for essential amino acids. **Journal of Nutrition**, v.127, p.1166-1175, 1997.

ROSTAGNO, R. S.; ALBINO, L. F. T.; DONZELE, J. L.; GOMES, P. G.; OLIVEIRA, R. F.; LOPES, D. C.; FERREIRA, A. S.; BARRETO, S. L. T.. E EUCLIDES, R. F. 2011. **Brazilian tables for poultry and pigs - feed composition and nutritional requirements**. 2. ed. Viçosa: Federal University of Viçosa, 2011. 141p.

SILVA, C. R.; GOMES, L. C.; BRANDÃO, F. R. Effect of feeding rate and frequency on tambaqui (*Colossoma macropomum*) growth, production and feeding costs during the first growth phase in cages. **Aquaculture**, v.264, p.135-139, 2007.

TANTIKITTI, C.; MARCH, B. E. Dynamics of plasma free amino acids in rainbow trout (*Oncorhynchus mykiss*) under variety of dietary conditions. **Fish Physiology Biochemistry**, v.14, p.179-194, 1995.

VIDAL JÚNIOR, M. V. **Techniques for determining the digestibility of nutrients in food for tambaqui (*Colossoma macropomum*)**. 2000. 96p. Thesis (Doctorate in Zootechnics) - Federal University of Viçosa, 2000.

YAMAMOTO, T.; SUGITA, T.; FURUITA, H. Essential amino acid supplementation to fish meal-based diets with low protein to energy ratios improves the protein utilisation in juvenile raimbow trout *Oncorhinchus mykiss*. **Aquaculture** v.246, n.1-4, p.379-391, 2005.

MIX
Papier aus verantwortungsvollen Quellen
Paper from responsible sources
FSC® C105338

Printed by Books on Demand GmbH, Norderstedt / Germany